FARM LIFE

by Elizabeth Spurr

illustrated by
Steve Björkman

SCHOLASTIC INC.

New York Toronto London Auckland Sydney
Mexico City New Delhi Hong Kong Buenos Aires

For Sherry and Ed Shahan—E. S.

For Michael, whose heart leans towards farms and ranches—S. B.

ISBN 0-439-65718-0

12 11 10 9 8 7 5 6 7 8 9/0

Printed in the U.S.A. 40

First Scholastic printing, March 2004

This artwork was painted with watercolors.

The text typeface is Kidprint and House-A-Rama Kingpin.

RED barn, **RED** barn, what are you keeping?

ONE rumbling tractor ready for reaping,

ONE baler, **ONE** sower,

ONE thresher, **ONE** mower.

This is the farm life—but only a part.

BLUE barn, **BLUE** barn, what are you hiding?

TWO sturdy stallions saddled for riding,

THREE fillies neighing,

FOUR hens a-laying.

This is the farm life—but only a part.

GREEN barn, **GREEN** barn,
what's in your loft?
FIVE bales of hay
with straw piled up soft,

SIX cribs of new corn,
SIX kittens newborn.
This is the farm life—
but only a part.

BROWN barn, **BROWN** barn,
what do you hold?
SEVEN small heifers
escaping the cold,

EIGHT ewes a-feeding,
EIGHT lambs a-bleating.
This is the farm life—but only a part.

GRAY barn, **GRAY** barn, what do you house?
NINE straw-filled stalls
with **NINE** sleepy sows,

TEN piglets suckling, Farmer Dan chuckling.
This is the farm life—but only a part.

WHITE house, **WHITE** house,
open your doors.

Show bright braided rugs on shiny oak floors,

Fresh hot tea steeping, down quilts for sleeping,

A crackling woodstove,
and a family to love.

This is the farm life—the whole and the heart.

GLOSSARY

baler: A machine that ties a field crop, such as hay, after it is gathered

down: Soft fluffy feathers

ewes: Female sheep

fillies: Young female horses

heifers: Young female cattle

loft: The second floor of a barn with an open inside wall

mower: A machine that cuts tall grasses and crops such as oats

piglet: A young pig

reaping: Cutting and gathering a ripened field crop

sow: A full-grown female pig

sower: A machine that scatters or plants seeds

stallion: A full-grown male horse

thresher: A machine that separates seeds from the plants

tractor: A vehicle used to pull a farm machine, such as a plow

Little Black Girl
Written by Vetta Shantell
Illustrated by
alia Edwards & Alexis Nelson

This book belongs to...

OZiOMARBJ.

Little Black Girl
did you know,
that your carbon
mixed with the sunlight
is the reason
why you glow,
why you grow,
and bloom like
a flower.

Little Black Girl did you know,
that even if you don't
hear it today, you are
BEAUTIFUL.

As beautiful as the trees on a
cloudy day, you let the world know
that even in the dark, something
beautiful is awake

Raise your head high
show those teeth,
because you my PRINCESS
are something to see!

Little Black Girl
id you know. that you create
your reality
so don't ever think low.

Think above and higher than
you can see.
because baby girl the sky is not
the limit to your
Destiny.

Beyond the sky there's a Galaxy,

the universe
made us all
and one day
you will see.

You will see how it all connects
and how it all makes sense,
But you must activate your Pineal gland
before you can live it

Your third eye is what helps
your mind reach to levels you
didn't know exist.
We must unlearn and then relearn,
that's the only way you'll get it!

I just want you to see,

that your people did not come from slavery.

Little
Black
Girl
I just want you
to know,

that your skin,
no matter the shade
is beautiful

Grow your own food,

Start your own business,

Don't ever think that age is the limit

So, always shoot for
the moon
because either way
you're among
the stars.

Take a deep breath,
meditate,
and relax your mind

Help your chakras
and energies
become more aligned.

Open your mind, research, and always
ask questions,
Let the truth set you free
so you can understand your personal legend.

Spend time with yourself, Write down your dreams,

because understanding yourself will give you the strength to always love genuinely.

If you believe it,
you can achieve it,
and so much
more!
So let's GO

make our dreams a reality
and accomplish what we have in store.

You are Inspirational

My desire for SUCCESS outweighs my fear of failure

To Love,
is
to Love
myself
FIRST. ♡

I am a young Goddess on a journey of personal growth

Everyday I will take
sometime to myself to
focus on things that will
benefit me and help
me achieve my
short-term and long-term
goals.

My **heart** is
filled with
love and strength
and I will not
let another
determine my
happiness

I will only think
positive
and will eliminate
all negative thoughts

It is only me
who can determine
my personal legend.

This book was written to inspire young women and families in our communities.

Black is **beautiful**.

Black is **love**.

Love yourself first and everything will fall into place.

Glossary

Aligned - place or arrange in a straight line.

Aura - a continuous change and flow of lifeand energy around ones body.

Carbon - A chemical element with the symbol "C" and the atomic number 6. This element has 6 electrons, 6 protons, and 6 neutrons. The reason for different shades of brown. (Another word for melanin. see melanin)

Chakras - pronounced: "cha"-"kra". A center of energy within the human body that helps regulate all its processes, from organ function to the immune system and emotions. Also known as the power of spirit and body energy. There are seven chakras throughout the body from the base of your spine to the crown of your head and each chakra has a specific color.

- Root Chakra - base of spine - color red
- Sacral Chakra - below the navel - color orange
- Solar Plexus - stomach - color yellow
- Heart - center of chest - color green
- Throat - base of throat - color blue
- Third Eye - forehead/between the eyes - color indigo
- Crown - top of head - color violet

Cosmos - the world or universe

Destiny - events that will necessarily happen to a particular person or thing in the future

Energies - combination of overlapping energy patterns that define an unique spiritual mental, emotional and physical aura of an individuals positive attention/vibes. see aura.

Esoteric - understood by only or meant for a small number of people with a specialized knowledge or interest.

Galaxy - a system of millions or even billions of stars, together with the gas and dust held together by gravitational attraction

Meditation - to clear your mind or to focus on one thought to achieve enlightenment. Used in an act to simply relax and gain greater understanding of life and oneself.

Melanin - provides some protection against skin damage from the sun and when exposed to the sun the melanocytes increase their production. A dark brown to black pigment occuring in the hair, skin and iris of the eye in people and animals. A dark biological pigment found in carbon. see carbon.

Melanocytes - produces melanin located in the skin

Melatonin - a hormone that is produced by the pineal gland that regulates sleep and wakefulness.

Pineal Gland - a small tissue in front of your brain and in the center of your forehead. This gland releases serotonin and melatonin and helps create dreams when your in deep sleep. An endocrine gland near the center of your brain. Also known as the third eye. (see third eye.)

Poverty - the condition in which you fall below the needed amount of resources to survive

Serotonin - is thought to regulate anxiety, happiness, and mood

Third Eye - esoteric concept of an "invisible eye" which provides perception beyond ordinary sight. (Also known as the gateway to higher conciousness) See esoteric.

Universe - all existing matter and space considered as a whole. Also known and the cosmos. The world of human experience. See cosmos.

Glossary